A Very Short Tour of the Mind

# *a* VERY SHORT TOUR *of the* MIND

*21 short walks around the human brain*

MICHAEL CORBALLIS

Duckworth Overlook

First published in the UK and US in 2013 by
Duckworth Overlook

LONDON
30 Calvin Street, London E1 6NW
Tel: 020 7490 7300
info@duckworth-publishers.co.uk
www.ducknet.co.uk
For bulk and special sales, please contact
sales@duckworth-publishers.co.uk
or write to us at the address above

NEW YORK
141 Wooster Street, New York, NY 10012
www.overlookpress.com

A catalogue record for this book is available
from the British Library

ISBNs
Hardback: 978-0-7156-4548-2
Mobipocket: 978-0-7156-4666-3
ePub: 978-0-7156-4667-0
Library PDF: 978-0-7156-4668-7

Manufactured in the UK

# Contents

1  Picking up the pieces  1
2  Swollen heads  6
3  On being upright  11
4  Why Italians gesticulate  16
5  Lost cousins  22
6  Attention!  26
7  On being right—or not  31
8  Split brain, split mind  36
9  About face  40
10  My oath  46
11  Small talk  51
12  Music  56
13  Remembrance of (some) things past  60
14  About time  65
15  Coloured days  69
16  I know what you're thinking  74
17  Mirror, mirror on the brain  78
18  Laughing matters  82
19  Telling left from right  87
20  The 10 per cent myth  92
21  Lies and bullshit  97
   Further reading  102
   Index  104

# 1

## *Picking up the pieces*

In 1891 my Irish great-grandfather, James Henry Corballis, published a book called *Forty-five Years of Sport*. It covered hunting, shooting, fishing, falconry and, as an afterthought, golf. I recommend it to anyone who wants to know how to mount a horse with a loaded gun, or where to place the golf ball in relation to one's feet before attempting to smite it along the fairway.

Forty-five years ago, in 1966, I took up my first position as a lecturer in psychology. Over the intervening years, I have taught at McGill University in Montreal, Canada, and the University of Auckland. I stopped actually teaching a couple of years ago, but I have remained in the sport—primarily as a researcher and writer, with the occasional lecture thrown in. I don't suppose I have learned or conveyed anything as useful as my great-grandfather did, but, well, it's been fun, as sport is supposed to be.

In the course of those forty-five years, I have seen rather dramatic changes in the nature of academic psychology. Scientific psychology began in the nineteenth century as the study of the mind. The main technique was introspection, turning the mind inwards to examine what might be there. Not much was discovered, though, probably because the mind does not have access to most of what it actually does—just as a car engine, say, does not itself understand how it works. Introspectionism gave way to behaviourism, a movement started by John B. Watson early in the twentieth century but developed later by B. F. Skinner. And when I came to psychology in the late 1950s behaviourism still ruled. Effectively, the concept of mind was abolished, and replaced with behaviour, the things people—and animals—actually do. Behaviour is directly observable and therefore amenable to measurement and scientific analysis. Behaviourists, though, saw no essential difference between humans and other animals, and psychological laboratories were filled with rats, and later with pigeons, which were considered more acquiescent and less likely to bite. My early experiences as a junior lecturer included making sure that rats were available and suitably placid for students in their laboratory classes. Even so, the odd student was bitten by the odd rat.

Soon, though, there came a rediscovery of the mind, in another revolution that saw the rats disappear, as though inveigled away by some Pied Piper. The pigeons, too, mostly flew away—although some do remain in some departments of psychology with an attachment to the past. Nevertheless, the cognitive revolution brought people back into the laboratory, largely replacing the rats and pigeons. It was heavily influenced by the emergence of digital computing, and by the linguistic theories of Noam Chomsky. The mind was reinvented as a computational device, although it was still studied by largely objective means—such as how quickly people respond to events, and how well they remember things.

Later, psychology discovered the brain, largely through the efforts of Donald O. Hebb, a distinguished Canadian psychologist, one of my mentors and later a senior colleague during my time at McGill. It turned out that through brain imaging and studying the effects of brain injury, we could look inside that large, wrinkled organ squeezed into our skulls to work out what different parts of the brain looked after—memory in the seahorse-shaped piece called the hippocampus, emotion right next door in the amygdala and

so on. And, by watching those regions in action while people looked at Jennifer Aniston or listened to Beethoven, we could begin to understand how the mind works. We humans may not have the largest brain in the animal kingdom, but we have proved to be the only animals capable of having a look inside.

The last decade or so has seen further dramatic changes—if not a revolution, then at least a vast broadening of methodology and subject matter. Some interest has again turned to animals, as we try to figure out how something as complex as the human mind could have evolved. Theories of brain function have become more elaborate, so psychology draws on brain science as well as on what people do. The Pied Piper who led the rats away, unlike his predecessor, brought children into the laboratory, so we could learn more about how their mental functions emerge. As the name implies, the cognitive revolution focused on thinking, neglecting emotion, but the new psychology is as concerned with feelings as it is with thought. Most importantly, the mind is now the focus of interdisciplinary study, the blending of information from diverse disciplines, including archaeologists, anthropologists, biologists, geneticists, linguists, neuroscientists, philosophers, as well as psychologists. It's enough to make the mind boggle.

In these twenty-one short walks, I have tried to convey something of the mosaic of the modern science of the mind. The topics were chosen much as the whim seized me. Many of them are adapted from pieces that were published as a column in *New Zealand Geographic*, whose word limit restricted them to bite-sized pieces, but enough, I hope, to convey a flavour—and I have embellished some of the original pieces a little. You may find some of them opinionated, but that's in the spirit of my great-grandfather, and of sport. I thank Margo White for suggesting the original column and James Frankham for agreeing to publish it. I am especially grateful to Sam Elworthy of Auckland University Press for his encouragement, enthusiasm and help over publication of the pieces in book form; and to Louise Belcher, Katrina Duncan and Anna Hodge for helping to make the book more attractive and readable. I also thank my wife, Barbara, for her tolerance while I wrote, but at least she had her golf, possibly inspired by *Forty-five Years of Sport*.

This book is especially dedicated to the three new ladies in my life: Lena, Natasha and Simone.

# 2

## *Swollen heads*

We humans are a swollen-headed lot. We like to think we're smarter than all other creatures, and perhaps uniquely blessed by some benevolent deity. Nevertheless, we need to be wary of our comfortable sense that we are at the top of the earthly hierarchy, since it provides a too-easy justification of the way we treat other animals. We eat them, kill them for sport, drink their milk, wear their skins, ride on their backs, ridicule them, house them in zoos and breed them to our own specifications.

How then do we justify our self-proclaimed superiority? One way is to appeal to our very swollen-headedness, looking to the brain itself as proof of our status on the planet. This strategy, though, has proven unexpectedly elusive. In terms of sheer brain size, for example, we have to defer to the elephant and the whale, whose brains are more than four times as big as our own. We cannot therefore claim to be the brainiest of creatures. Perhaps, though, the absolute size of the brain is not really a good measure of intelligence. Large animals need large brains simply to control those big bodies, and deal with all of the information that arrives from their large surfaces. So maybe we should consider not the absolute size of the brain, but rather the ratio of brain size to body size.

Here we come out rather better. Our brains weigh about 2.1 per cent of our body weights, while those of our closest cousins, the chimpanzee and bonobo, are about 0.61 per cent and 0.69 per cent, respectively. The figures for the Asian elephant and killer whale are 0.15 per cent and 0.094 per cent, respectively. So far so good—we can dismiss those lofty elephants and voluminous whales as large but fairly dumb. Sadly, though, the mouse comes out better than we do, with 3.2 per cent, and in small birds the ratio may be as high as 8 per cent.

One approach to this problem is to retreat into mathematics, and hope to bamboozle our creature cousins with equations. Other things being equal, smaller animals have larger ratios than bigger animals do. The psychologist Harry Jerison plotted log brain size against log body size across a wide range of species, and then used a technique called linear regression to compute the slope of the line relating one measure to the other. The slope of this line was ⅔, meaning that body size mattered less the larger the animal. This led to calculation of the expected brain size based on body size, and dividing this into the brain weight leads to what Jerison called the encephalization quotient.*

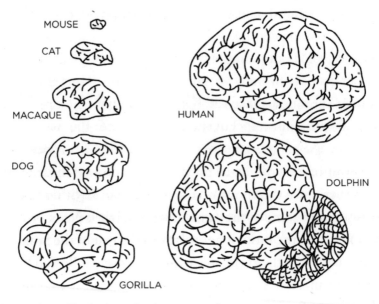

MOUSE

CAT

MACAQUE

HUMAN

DOG

DOLPHIN

GORILLA

Comparison of brain sizes of various mammals

---

*   For those who need to know, the expected brain size for an animal whose body weight is w is $0.12w^{\frac{2}{3}}$. Dividing this into the actual brain weight gives the encephalization quotient.

This quotient turned out to be 7.44 in humans, followed by dolphins at 5.31 and chimpanzees at 2.49. The elephant weighs in, as it were, with a quotient of 1.87 and rats have a miserly quotient of 0.4. The mouse is blessedly reduced to a quotient of about 0.5, so we can stop worrying about that. This quotient also pretty well gets rid of any serious challenge from birds, although comparison is difficult because the slope of the line is rather different for birds. We should be wary of the dolphin, though, which may be closer to us than we'd like to think, and of course there may be other creatures busily working on formulae to prove that they, after all, are top dogs.

Another approach is to examine the neocortex, the outer layer of the brain which houses higher-order functions such as language, thinking and memory. Focusing on the neocortex has the decided advantage of getting rid of birds altogether, since they possess no neocortex at all (other brain areas perform some of the same functions, but let's not go there). Taking the ratio of neocortex to the rest of the brain places us above other primates with 4.1, closely followed by the chimpanzee at 3.2, gorilla at 2.65, orangutan at 2.99 and gibbon at 2.08.

This ratio seems to increase with group size—gibbons hang out in groups of about fifteen, orangutans in groups of about fifty and chimpanzees in groups of about sixty-five. The rather solitary gorilla seems to be something of an exception. Based on what we know from the brain sizes of our hominin forebears, group size should have been roughly constant at around sixty in the australopithecines, but increased steadily with the emergence of the genus *Homo* from around 2.5 million years ago, culminating in *Homo sapiens*. According to the formula, humans should belong in groups of about 148, which is roughly the typical size of Neolithic villages. Of course, modern cities contain vastly more than that, but if you add up the people you are actually acquainted with it may be not too far off the mark. The relation of neocortical size to group size might be taken to mean that neocortical evolution, and perhaps intelligence itself, was driven by the pressures of social interaction.

If the path toward demonstrating our superior brain power has seemed tortuous, we can perhaps at least gain comfort from the thought that we are the only species working on the problem—or so it seems.

# 3

## *On being upright*

Humans are unique among the primates (monkeys and apes) in being habitually bipedal. Chimpanzees can stand and even stagger around a bit on two legs, but their regular means of getting around is knuckle walking, in which the forearms serve as forelegs, with the knuckles touching the ground. Since the chimpanzee (along with the bonobo) is our closest relative, it has been commonly assumed that we too once were knuckle-walkers.

The recent discovery of a near-complete fossil called *Ardipithecus ramidus*, popularly known as 'Ardi', challenges this idea. Ardi dates from some five million years ago, close to the point at which our forebears diverged from the line leading to modern chimpanzees and bonobos. She appears not to have been a knuckle-walker, but stood upright. Her foot was nevertheless shaped for grasping, suggesting that she was still primarily a tree-dweller. Her hand was closer to that of a modern human than to that of chimpanzee or bonobo. Some have concluded that bipedalism, at least as a posture for manoeuvring in the forest canopy, may actually date from tens of millions of years ago, and that knuckle-walking in gorillas and chimpanzees emerged after the split from the human line.

Our earliest human ancestors evidently did walk about in bipedal fashion, albeit clumsily, but the effortless bipedal gait we now enjoy probably came about much later, when the genus *Homo* emerged around 2.5 million years ago. (Our own species, *Homo sapiens*, emerged only within the last 200,000 years.) In some respects, though, bipedalism seems not to have been a good idea—as the rampant pigs in George Orwell's *Animal Farm* put it: 'Four legs good, two legs bad!' Adapting to our two-legged stance may be ultimately responsible for back and neck problems, haemorrhoids, hernias, and the excessive pain of giving birth. Sciatica can be blamed on the fact that the spinal nerves and spinal discs are too close together, again a result of a bipedal adaptation. Bipedal walking is not even especially efficient—a knuckle-walking chimp is said to be able to reach speeds of up to 48 kilometres per hour, whereas a top athlete can manage only about 30.

Because of the reshaping of the pelvis for walking, human babies are born prematurely, at least by primate standards. According to the normal primate gestation period, they should be born at about eighteen months but, as any mother can tell, this would make an already difficult process physically impossible. Early birth means that babies take a long time to learn to walk, and they crawl on four limbs before walking. They are therefore especially vulnerable to predation, as we are reminded by the incident in Central Australia when Lindy Chamberlain's baby daughter was carried off by a dingo.

But there must have been some advantage to being bipedal that outweighed these impediments. For one thing, it freed the hands. As our forebears moved from forested areas to the more open savannah, the hands no doubt proved useful in scavenging and carrying foodstuffs back to the group. Soon, prompted no doubt by the enlarging brain, the hands were put to use in developing tools, first for cutting and scraping carcasses, and then for developing hunting weapons, such as knives, spears or fish-hooks. The freeing of the hands accompanied by an upright stance created an enhanced ability to throw, either to ward off predators or kill prey.

Humans are vastly superior to other primates in our ability to throw objects with accuracy and force. Even if that ability has atrophied in our modern sedentary lives, it is apparent in sports such as cricket or baseball. The freeing of the hands may even be responsible for language, but we'll tackle that story later.

Indeed, in the overall reckoning, we can probably thank bipedalism for the dominance of humans over other primates. We share nearly 99 per cent of our deoxyribonucleic acid (DNA) with the chimpanzee, but humans have come to dominate the planet, whereas chimps and other apes are confined to ever-diminishing environments, and are threatened with imminent extinction. We have somehow turned the disadvantages of bipedalism to great advantage. Although the size of the birth canal restricts the size of the human brain at birth, prematurity actually allows the human brain to grow large after birth, where it is no longer constrained by the size of the womb. As we saw earlier, the human brain is about three times as big as expected in a primate of our size. Moreover, the brain is at its most impressionable while it is growing, and early birth means that the infant brain is exposed to the social environment during periods of growth that are critical for acquiring language. And the helplessness of babies means that they are given extra nurture, essential for social bonding and empathy.

The spiral of manipulation, increasing intelligence and effective communication, all stemming from bipedalism, makes us what we are today—a dominant, manipulative, dangerous species, albeit one capable of acts of altruism and goodwill that may save us from self-immolation. We might ponder the fact that there have been some twenty hominin species identified as having existed since the split from the chimpanzee line, but only one remains—us. Nevertheless, the present-day plight of our nearest relatives, the chimpanzees and bonobos, is altogether more dire.

Be thankful, but perhaps also feel a little guilty, that we perservered with our two-leggedness and eventually achieved dominance over our knuckle-walking cousins.

# 4

## *Why Italians gesticulate*

Most linguists and cognitive scientists agree that we humans are uniquely blessed with the gift of language. Compared to human speech, animal calls are stereotyped and fixed, tied to specific situations such as mating, territorial claims, expressing aggression or raising alarm. The nearest equivalent to speech seems to come not from our nearest relatives, the great apes, but from birds. Parrots, for example, can imitate human speech, and even be taught to answer simple questions, like naming the colour of a block or counting the number of objects in a display (but only up to about six). Songbirds in the wild generate complex calls, but these are largely repetitive and probably serve primarily as identification codes. They have none of the individual variety that enables humans to express a virtually infinite range of ideas, thoughts and opinions.

For evolutionary theorists, the uniqueness of human language poses something of a problem. The seventeenth-century French philosopher René Descartes thought language so special that it must be a gift from God, and even materialistic scientists have sometimes supposed that there must have been some miraculous event—a fortuitous genetic mutation, perhaps—that gave us the gift of the gab.

My own solution—although not everyone agrees—is that language evolved, not from animal calls, but from manual gestures. Our great-ape cousins actually possess a much better range of manual gestures than of vocal sounds, and gesture naturally to each other in the wild. It is becoming clear, moreover, that their gestures are more deliberate and conversation-like than are their vocalisations. Attempts to teach captive chimps and bonobos to speak have been spectacularly unsuccessful, but some real progress has been made toward teaching them to communicate in a simplified form of sign language, or by pointing to symbols on a specially designed keyboard. It's not true language, but it's much more language-like than their attempts at speech.

We saw in the previous piece that our hominin forebears stood up on two legs and waved goodbye to the apes. This freed their hands from locomotory duty and, among other advantages, would have enhanced their ability to communicate manually. From over two million years ago, too, our industrious ancestors began to make stone tools, and this may have led to mimed communication about how tools work, and how to make them. Along with the development of tools, brain size increased dramatically. These developments probably signalled the emergence of a more elaborate group structure and complex communication, culminating in language, with its complexities of grammar and capacity to generate unlimited meanings.

If the gestural theory is correct, language must have switched from a manual to a vocal point at some stage in our evolution. This probably required further adaptations, including a lowering of the larynx in the neck, and a flattening of the face and shortening of the tongue. Fossil evidence suggests that these features were absent, or at least incomplete, even in the Neanderthals, depriving them of articulate speech. Jean Auel, in her novel *The Clan of the Cave Bear*, tells how Ayla, a Cro-Magnon girl, is orphaned after an earthquake and adopted by Neanderthals, who are capable of only limited speech but have a highly developed sign language. I like to think she is right about this. The Neanderthals had brains as least as large as our own, but died out some 30,000 years ago. We voluble humans may have somehow talked them out of existence.

So was this the terrible secret of *sapiens*? My guess is that it was not superior strength, or even superior brain power, that won the day. Rather, it was the late transition from gestural language to speech. So why should this transition have been so decisive—and indeed so destructive? Clearly, the advantages of speech over gesture were not linguistic, since sign languages are as linguistically sophisticated as vocal ones. The advantages were probably practical. Speech allows communication at night, or when the line of sight is impeded. Speech requires much less energy, since it effectively piggybacks on breathing, which we do anyway. Signing, in contrast, can be exhausting—instructors in sign-language courses, I'm told, often need regular massage to cope with the sheer physical demands.

Most importantly, though, speech allowed a second freeing of the hands. The retreat of communication into the mouth is an early example of miniaturisation, allowing the hands, and indeed the rest of the body, to be put to use in other activities, such as making and using tools, or carrying possessions around as our restless forebears moved from one location to another. This may account for the extraordinary explosion of technology (including weapons), art, bodily decoration and sheer cultural diversity that characterise our species. It also allowed our talkative forebears to explain manual skills while at the same time demonstrating them, as in modern cooking shows on television. These effects were no doubt cumulative, but in the course of time the gesturing Neanderthals never stood a chance.

Our technological advances, though, may have captured our hands once again, as we tap on our laptops, point laser-emitting devices at our televisions or tweet digitally on our cellphones. Perhaps we will need to find other ways to communicate with our devices, maybe by talking to them, or even through hooking up to direct measurement of brain activity. But whatever we do, our restless hands will no doubt always find new ways to occupy themselves—and we will continue to miniaturise, so that as fast as we find new things for our hands to do, we will find ways to relieve them again.

You may think it an odd idea that language was originally based on manual gestures, but note that deaf children can acquire sign languages just as easily and naturally as the rest of us learn to speak. Deaf kids even babble in sign.

At Gallaudet University in Washington, DC, the official language is American Sign Language, and students study all the usual subjects—even poetry. And of course we all—but especially the Italians—gesture with our hands as we speak. The gestural origins of language are all around us.

# 5

## *Lost cousins*

In 1856 the remains of a prehistoric, human-like figure were discovered in the Neander valley in Germany, causing some consternation in polite European society. The Canadian geologist John William Dawson greeted the discovery as follows:

> It may have been one of those wild men, half-crazed, half-idiotic, cruel and strong, who are always more or less to be found living on the outskirts of barbarous tribes, and who now and then appear in civilized communities to be consigned perhaps to the penitentiary or the gallows, when their murderous propensities manifest themselves.

A more recent commentator wrote that the Neanderthal 'came into the world of the Victorians like a naked savage into a ladies' sewing circle'.

Analysis of deoxyribonucleic acid (DNA) extracted from Neanderthals now reveals that we share with them a common ancestry dating from around half a million years ago, and that they became extinct around 30,000 years ago in Europe, some 20,000 years after early humans arrived there. The uncomfortable closeness of the Neanderthals to ourselves seems to provide a strong imperative to insist that they were a separate species, no doubt less intelligent and more brutish. Though their brains were slightly larger than ours, we can attribute this to their slightly larger bodies, allowing us to believe that we humans are just as intellectually able, if not more so. As I have mentioned, modern scholars have argued that the Neanderthals didn't even have the blessing of language, on the grounds that their vocal tracts were not fully adapted for speech. But speech should not be identified with language and the Neanderthals may well have communicated effectively with gestures.

In 2010 an international consortium of scientists published a draft sequence of Neanderthal DNA indicating that we are even closer to our supposedly brutish cousins than previously thought. Comparing this DNA with our own reveals that the Neanderthals contributed some 1 to 4 per cent of the genome of modern humans. This suggests that the early humans who came out of Africa some 65,000 years ago went on a spree of rape and pillage; once they arrived in Europe they set about slaughtering the hapless Neanderthals but occasionally also mating with them. Well, perhaps it wasn't quite as bad as that, since the two groups coexisted in Europe for some 20,000 years, but the scenario is uncomfortably reminiscent of the way our species carries on when one bunch invades the territory of another. And, of course, that shared DNA blurs the notion that the Neanderthals were a separate species.

It gets worse. DNA extracted from the little finger bone of another fossil discovered in the Denisova Cave in southern Siberia suggests that there was another human-like species around, neither human nor Neanderthal, dating from some 30,000 to 50,000 years ago. Comparing that DNA with modern human DNA suggests that some of our human ancestors also mated with the Denisovans, who contributed about a twentieth of the genome of Melanesians who now inhabit Papua New Guinea and the islands northeast of Australia.

The purest 'humans' must therefore be those Africans who did not join the exodus of 60,000 or so years ago. Those who did head out from Africa appear to have interbred, albeit to a limited extent, with those who had migrated much earlier. It will not be surprising if new discoveries reveal further skeletons in the cupboard.

These recent revelations might seem to support ideas of racial differences between contemporary groups, perhaps also suggesting genetic differences in intelligence or social refinement. However, despite the donations of small amounts of Neanderthal and Denisovan DNA, we are all basically of African stock. Our African forebears seem to have evolved qualities that enabled them to eventually populate the globe, while the Neanderthals and Denisovans perished. Those qualities may well include superior technology, enhanced communication and even bonding rituals such as religion and rugby. In the light of human history, though, we should reflect that a major part of technological advance was the development of increasingly powerful weapons, and that rape and pillage have featured as prominently as the ties of human generosity and altruism.

And, with respect to the Neanderthals and Denisovans, it's too late to say we're sorry.

# 6

## *Attention!*

We live in a world of bewildering complexity. There is just far too much going on for us to be aware of all the sights, sounds, smells, tastes and touches that constantly harass our senses. Our brains have therefore evolved to select some aspects of the world for mental processing and ignore the rest. That process is what we call attention.

Its selective nature is easily demonstrated. In one famous example, people are shown a video clip of a basketball game, and are asked to count how many times the ball is passed by the players. People are generally so intent on watching the players and counting the passes that they fail to notice a person dressed in a gorilla suit walking through the scene. Similarly, if you play two different messages simultaneously into each ear via headphones and ask people to listen to one of the messages, they generally fail to pick up any information from the other, even though it's equally loud. This proves that attention can be operated internally, by the mind, and not simply by orienting the ears. Visual attention, though, generally depends on where the eyes are looking. Even so, it is possible to look straight ahead and yet pay attention to events out of the corner of the eye—a device that can be used by rugby or netball players to trick the opposition. Or by school teachers, anxious to detect mischief-makers.

Nature has equipped us with automatic mechanisms to capture events that might be important for survival. Loud noises, sudden movements, brilliant flashes of light— these all divert us from what we are doing, in case they signal danger. So does extreme pain. A severe toothache, for example, may divert us from reading an assignment— although equally we might indulge in some especially engaging activity, such as listening to a favourite band, to take the mind off the pain. Alarm systems are typically loud and jarring, although we may also be tuned to more subtle events. A mother may be especially alert to the sound of her baby crying, even from a distance, and all of us are sensitised to the sound of our own names being spoken, even when whispered.

27

We are not slaves to the environment, though; attention can be controlled voluntarily, as when we choose to read a book, listen to a lecture or a piece of music, or solve a crossword puzzle. To a degree, then, we can filter out most environmental distractions, although not all. In his poem 'The Canonization', poet John Donne famously exclaimed 'For God's sake hold your tongue, and let me love'. Attention is not always directed to the external world. We can beam it inwards, as when we are lost in thought or reverie.

Attention requires a fine balance between concentration on the task at hand and awareness of the surrounding environment. We cannot be so intent on solving a Sudoku puzzle that we fail to observe the conflagration around us, nor can we be so distractible that we fail to complete any task requiring sustained concentration. Sometimes the balance is disturbed. People with damage to the frontal lobes often have difficulty shifting from one task to another, or in adapting to changes in instructions. This suggests that the frontal lobes are the drivers of attention. People with attention deficit disorder have the opposite problem, they are too easily distracted. In general, boys seem more prone to this than are girls and there may be an evolutionary reason for this. In a hunter-gatherer society, it was probably the males who did the hunting, and hunters should pay attention to all distractions—that noise to the left may be a hungry lion.

An example of the effect of hemineglect in a sketch by Lovis Corinth

The human brain is curiously asymmetrical in the way it controls attention. The left brain attends to the right side of space; the right to both sides, albeit with some bias toward the left. Damage to the right side may therefore cause a patient to lose awareness of events on the left, a phenomenon known as hemineglect. The patient may eat from only the right side of their plate, dress only the right side of their body, ignore those who address them from the left, and are easily beaten at chess by attacking on the left flank. A famous example is the German artist Lovis Corinth, who suffered a right-brain stroke in 1911 but continued to draw and paint until his death fourteen years later. Much of his work shows a neglect of the left side, as in the portrait above.

Just why the brain should function in this asymmetrical way is not clear. Perhaps it's because in most of us the left side of the brain is largely taken up with language, and so loses some of its capacity to direct attention to space. Left-brain damage seldom results in neglect of the right side, or does so only transiently, and there is no evidence that non-human animals show a similar asymmetry. We are the lopsided ape.

# 7

## *On being right—or not*

*Cack-handed, cow-pawed, dolly-pawed, ker-handed, left-plug, southpaw, squiffy* . . . are just some of the names that have been applied to left-handers. They have an air of insult, implying that lefties are somehow inferior, or at least odd. On the other hand, as it were, we right-handers are, well, just *right*. Our language betrays the distinction in other ways too; *gauche* and *sinister* originally meant 'left', while *adroit* and *dexterous* have an agreeable air of rightness, if not righteousness.

The plight of the left-handed derives mainly from the fact that lefties make up only about a tenth of the population, and so fall victim to the tyranny of the right-handed majority.

Any clumsiness can generally be attributed to the fact that the manufactured world is designed for the right-handed. Asymmetrical objects such as scissors and corkscrews conspire to frustrate the left-hander, as do the placement of door handles or the ordering of pages in books and magazines. We righties just don't make it easy for lefties when we force them to shake hands or salute with the right hand, and in a bygone age left-handers were forced to write or eat with the right hand, often with unhappy consequences.

Perhaps to compensate for discrimination against lefties, though, a romantic view of left-handers sees them as more creative or artistic than right-handers, and this sometimes leads to false claims. For example, Wikipedia once listed Albert Einstein, Benjamin Franklin, Pablo Picasso and Leonardo da Vinci among famous scientists and artists who were left-handed. The first three were almost certainly right-handed. Wikipedia was right about da Vinci, but should have left the others out.

The proportion of natural left-handers in the population is roughly constant across different cultures, and has probably been that way for thousands of years. Just why this is so is an enduring mystery. Left-handedness does tend to run in families, but is only weakly inherited. Having one left-handed parent raises the chances of being left-handed from just under 10 per cent to about 20 per cent, and having two left-handed parents raises it further to about 26 per cent. Left-handedness, then, does not breed true, but there does seem to be a genetic influence.

There is as yet no convincing evidence of the responsible gene or genes, but the best guess is that the genetic influence is not on whether an individual is right- or left-handed, but rather on whether an individual is right-handed or not. Left-handers, then, are not so much left-handed as lacking any genetic disposition toward right-handedness. In these individuals, handedness is a matter of chance—some turn out left-handed, some right-handed and a small minority are mixed-handed, or ambidextrous. This theory explains why left-handers are mixed with respect to other asymmetries, such as eye dominance, or which side of the brain is dominant for language. It also helps explain why left-handedness does not breed true.

People lacking overall handedness often excel in sports—one thinks of cricketers who bat left-handed and bowl right-handed (or vice versa), or balanced rugby or soccer players who kick equally well with either foot. Yet mixed-handers may also have a slightly higher risk of disorders in skills that depend strongly on brain asymmetry for optimal efficiency. One such skill is language, and those without consistent handedness appear to be slightly more at risk for disorders such as stuttering or reading disability. A large-scale study of eleven-year-olds in the United Kingdom showed that mixed-handers scored slightly worse than either lefties or righties on various tests of academic achievement, and our own analyses of data drawn from the television programme *Test the Nation: The New Zealand IQ Test*, aired in 2003, showed the same trend.

In evolutionary terms, selection may have conspired to hold tendencies to symmetry and asymmetry in balance. We belong to an ancient but encompassing phylum known as the Bilateria, with a body plan that is fundamentally bilaterally symmetrical. This is no doubt an adaptation to life on a planet where events impinge without obvious bias toward one or other side. There are some respects, though, in which bilateral symmetry is an impediment. A useful analogy is the motor car, which is largely symmetrical with respect to its outward shape. It would be inefficient, though, to require that its engine and internal parts also be symmetrical, such that the left side was exactly mirrored in the right side. Similarly, the internal workings of the brain could be impeded if nature was to insist on symmetry. This is especially true of brain processes, such as those involved in language, that have little to do with the spatial world. Consequently asymmetries have evolved, typically allowing one side of the brain to take control. But too much asymmetry might have us forever moving in circles, or vulnerable to predators who sneak up on the weaker side.

Curiously, studies have also shown that mixed-handers score higher on what has been called magical ideation—a tendency to superstition and irrational belief. One wonders, then, if the asymmetrical brain might be more hospitable to rationality, while a symmetrical brain might accommodate the more spiritual, religious side of human nature. Could it be, then, that handedness holds the key to that most fundamental of human dichotomies, science versus religion?

# 8

## *Split brain, split mind*

Suppose your brain were cut in half, separating the left and right sides. Would your mind also be divided? This question has been of philosophical interest since René Descartes, the seventeenth-century French philosopher, formulated mind–body dualism—the notion that the mind is distinct from the body, and therefore from the brain. Early in the twentieth century William McDougall, a prominent psychologist, wanted to test this doctrine by requesting that his own brain be split should he ever become incurably ill. He was a dualist, believing that his over-arching mind would unite the physically divided halves of his brain. But the surgery was never performed, and McDougall died in 1938. One might still wonder, I suppose, whether his mind has soldiered on.

Split-brain surgery was actually performed, though, in the 1960s, when two Californian neurosurgeons, Philip J. Vogel and Joseph E. Bogen, attempted to provide relief to patients with intractable epilepsy. To do so, they decided to cut each patient's corpus callosum, the main band of fibres connecting the two sides of the brain. The idea was to prevent the spread of seizures, and in this respect the operation was more successful than anticipated. The seizures were either greatly reduced or readily brought under control with drug therapy. At first, the operation seemed to confirm dualism. Bogen remarked that these split-brain patients showed what he called 'social ordinariness', meaning that their everyday behaviour seemed entirely normal. One patient, asked how he felt after the operation, joked that he felt fine—except for a splitting headache.

Dualism, though, was effectively discounted when psychologist Roger W. Sperry undertook more subtle experiments to test each side of the brain separately, and came to the conclusion that each side dwelt in its own individual consciousness, with its own thoughts, feelings and memories. He found, moreover, that each side seemed to exhibit a different kind of consciousness. Only the left side could produce speech, or do calculations, while the right side seemed more proficient at spatial or emotional processing. Sperry received the Nobel Prize for his work in 1981.

The split brain also seemed to capture the spirit of the divisive 1970s. Bogen, along with journalist Robert E. Ornstein, set about popularising the idea of the brain having two sides, declaring the left to be the side of rationality, logic and masculine thought, and the right the side of creativity, intuition and femininity. The distinction rapidly became part of folklore, leading to what might be termed a right-brain cult, fed in part by feminism and resistance to the Vietnam War, with the left brain embodying the military-industrial establishment of the West and the right brain the supposedly peace-loving East. The 1970s slogan 'Make love, not war' was a call to arms—embracing ones, not military ones—to the right brain.

All this was a gross exaggeration of the neurological facts, but the myth persists. A Google search for 'right brain' still produces some eighty million entries, many of them calling on techniques to encourage right-brain activity in art, education, therapy, business and even literature, despite the left brain being responsible for language. The right brain has become not so much an emblem of peace and love as a means of profit-making. Beware of sellers of right-brain potions.

In the meantime, split-brain research has dwindled, in part because the operation is now seldom performed, except in some extreme cases in which psychological testing is difficult, if not impossible. Drug therapies for epilepsy have improved to the point that surgery is either unnecessary, or is directed more precisely to the locus of the problem. Research with the few remaining testable patients has also shown that the split brain is not quite so split as originally thought.

Although the corpus callosum connects the two cerebral cortices, which house higher mental processes, there are lower connections that maintain a unity of emotion and probably also consciousness. Much of the early work was based on the fact that vision is neatly divided down the middle, with the right side of space projecting to the left side of the brain and the left side of space to the right side of the brain. It now transpires that the visual world is at least partially connected via subcortical pathways, and one split-brained patient for years drove his ute on the highways of the northeastern United States without incident.

Split-brain research has nevertheless taught us a lot about how the mind works, and now takes its place alongside brain imaging and evidence from other kinds of neurological interventions. It may never be possible to entirely prove or disprove the theory that the mind is separate from the brain—opinion remains, you might say, split—but the more we understand the brain and observe its influence on how we think, feel and behave, the more unlikely that theory becomes.

# 9

## *About face*

How many faces do you know? Hundreds? Thousands?
More than you realise, I bet. Consider first your relatives,
schoolmates, workmates, neighbours, casual friends. Then
there are celebrities, such as film stars, rugby players, tennis
stars, even historical figures. Here is a random list: Henry
VIII, Beethoven, Nelson Mandela, Winston Churchill,
Gregory Peck, Marilyn Monroe, John Lennon, Björn Borg,
Salman Rushdie, Princess Diana, Brad Pitt, David Beckham.
No doubt I betray my age and cultural habits, but I can
picture each one of them. And there are people you simply
see around a lot, even though you don't really know them,
but you'd probably recognise them if you unexpectedly saw
them on the street in a foreign city.

We also recognise friends and acquaintances after long lapses of time, during which their faces have changed. Hair might have diminished or changed colour, wrinkles increased, chins multiplied, but somehow the same old person shines through. We recognise people whether full-face or in profile, in colour or black and white, laughing or frowning. Just as impressive is our ability to know that we have *not* seen a particular face before. Walk the streets of London, say, and you might pass hundreds of people and know that you don't know them, and yet they are all distinctive in their own ways.

These feats are all the more remarkable in that people's faces don't actually differ very much. They have two eyes, a nose and a mouth, similarly located on everyone. How can we distinguish, let alone remember, objects that are so alike? Their names don't seem to be involved. We often remember a face but don't recall the name, or perhaps don't even know it, but it's rare to remember someone's name but then not recognise them when we see them.

Somehow, then, faces are 'special', critical to our sense of belonging and well-being. For this reason, perhaps, a special part of the brain in the temporal lobe, known as the fusiform face area (FFA), has been assigned the task of encoding information about faces, and providing quick information as to whether we know someone or not. Injury to that part of the brain can result in a specific loss of the ability to recognise faces, while recognition of other objects remains intact. This is known as prosopagnosia, vividly described in Oliver Sacks's book *The Man Who Mistook His Wife for a Hat.*

In one remarkable study, tiny electrodes were inserted into the temporal lobes of patients to record the activity of single cells (or neurons) from among the 100 billion which make up the bulk of the human brain. The purpose of this was to locate the source of epileptic seizures. In one patient, one particular cell was activated whenever the patient was shown pictures containing the face of the actress Jennifer Aniston. It responded to a wide variety of her poses, but was knowingly silent when Brad Pitt appeared in the same picture. (Aniston and Pitt had recently divorced after five years of marriage.)

As though to compensate for ability to recognise faces under different conditions, there are some transformations which greatly impede recognition. Faces are peculiarly difficult to recognise upside down, unlike most other common objects, such as a bicycle or a tree. Figure 1 depicts a face, but it's almost impossible to see it as a face unless you turn it around—and even then you might have to work on it a bit. Figure 2 looks recognisable as a smiling Margaret Thatcher, but turn it round and you're in for a rude shock. The trick here was to leave the mouth and eyes the right way round but turn the rest of the face upside down. This seems to show that eyes and mouth are fairly critical to recognition, but only when viewed the right way up.

Figure 1

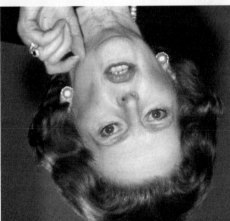

Figure 2

Figure 3

Faces, unlike most objects, are almost unrecognisable as photographic negatives. This can be demonstrated with a neat trick. Stare at the four dots in the middle of figure 3 for 20 seconds or so, and then look at a blank wall or sheet of paper. The result is known as a negative afterimage, and in this case you should then see, and perhaps recognise, the face of which figure 3 is the unrecognisable negative.

There is a story, probably apocryphal, of two Oxford professors, who went bathing in the nude in the local river. While drying themselves on the river bank, a boat full of female undergraduates came by. Professor A immediately covered his lower body with his towel, but Professor B covered his face. 'Why did you do that?' asked Professor A. 'Most people,' replied Professor B, 'recognise me by my face.'

# 10

## *My oath*

We all swear, although some people swear they don't. In one sense, swearing is benign, as in swearing on the Bible in court, or swearing by something you believe in, like a cure for warts. But we also swear in the more interesting and degenerate sense of using taboo words, of the sort we are urged not to utter in front of the kids.

And kids, too, are often reprimanded or punished for using taboo words. The wonder is how they learned them in the first place. Perhaps they are spread by older siblings or adult moles, or by parents who simply can't refrain even when the kids are around. Children anyway seem to have an instinctive reaction to overly moralising parents or teachers, and irresistibly seize the opportunity to shock. Being told one must never say a particular word creates an irrepressible urge to say it, just as being told one must not think of a polka-dotted elephant makes it very hard not to create precisely that image in the mind.

Surveys show that about two-thirds of swearing has to do with frustration, anger or surprise. As an aggressive weapon, swearing may serve as a substitute for physical assault. It is more common in those who rank low in social standing, in extroverts, in those with high levels of hostility, and less common in people who are agreeable, conscientious, religious or sexually anxious. Swearing can also be simply a kind of slang, often no more offensive than other cult words, such as 'awesome' or 'cool'. It may also be a badge of allegiance, especially among groups of men.

Swearing in anger seems to depend on deep areas of the brain, including a structure known as the amygdala, which is involved in emotion. But it is perhaps also a question of whether higher-order structures, and especially the frontal lobes, can suppress the impulse to swear. People who have lost the normal use of speech due to cortical damage often retain the singular ability to utter profanities, and understandably may do so profusely. This suggests that swearing can be automatic and ungoverned. Another example comes from Tourette's syndrome, an inherited neurological condition characterised by involuntary swearing and cursing (known as coprolalia), along with tics, head jerking, spitting and yelping. Victims of the disorder are embarrassed by their profanities, but have no control over them. Somehow, the balance between the well-mannered cortex and more primitive emotional centres in the brain has been disturbed.

In the recent movie *The King's Speech*, George VI, a chronic stutterer, is depicted as frequently swearing. While this may have been simply a reaction to frustration, his swearing is relatively fluent, and the speech therapist is shown using swearing as a means of controlling the King's stutter during his public speeches. I am told though that Peter Conradi, co-author of the book on which the movie was based, swears this didn't actually happen.

The earliest profanities are probably religious, prompted by strict teaching not to take the names of God or gods in vain. In an increasingly secular society, especially in the West, these taboos have largely lost their power to offend. Oaths like 'Jesus', 'Christ' or 'God in heaven' are now almost empty of shock value. In earlier times, religious profanities were often sanitised, as in 'jeepers', 'cripes' or 'gosh', but such terms have largely disappeared. In some cultures, though, religious terms have retained the power to shock. In French

Quebec, the most effective profanities still relate to the Church and its liturgy, as in expletives such as 'tabarnak' (tabernacle) or 'calice' (chalice). 'Merde' is relatively mild.

Nevertheless, religious doctrine may also underlie the taboos against terms relating to defecation and sexual function, but these too have lost much of their impact, although British actor Stephen Fry, recently stuck in a lift, expressed his frustration by tweeting 'Arse, poo and widdle' to his many followers. Expressions like 'bugger', 'shit', 'piss off', 'bollocks' or 'cock-up' punctuate normal conversation without arousing much disapproval, except perhaps in some leafy avenues of genteel society. The F-word was once replaced by sanitised variants, such as 'frigging' or 'fecking', but is now part of regular discourse and is heard widely on television, often as a culinary accompaniment. There are wide cultural differences, of course, and even in the United States such profanities on public television are much less acceptable than in the United Kingdom or feckless Australasia. The C-word is still largely taboo, perhaps due to a persisting sexual taboo mixed with the residues of old-time chivalry—and maybe a pinch of feminism.

Change can be swift. In 1914 the phrase 'Not bloody likely!' caused an uproar when uttered on an English stage, in 1956 All Black Peter Jones shocked the nation by declaring himself 'absolutely buggered' on public radio, and in 1972 Germaine Greer was fined for saying 'bullshit' in a speech at the Auckland Town Hall. Domains of disapproval have now largely shifted from the religious and scatological to the political, probably because we are controlled by lawmakers rather than religious authorities. The strongest taboos are against racial terms. The N-word, for example, is rarely heard on television, but may surface in racist insults hurled by white supremacist skinheads. Other derogatory terms for cultural and indigenous minorities have largely disappeared from everyday discourse, and retain a genuine power to offend.

Taboo words will persist, if only because taboos themselves reflect necessary controls over human behaviour. Unchecked, we are an unruly species. But taboo words are also valuable in providing outlets for frustration that fall short of physical violence. A world without taboos, and taboo words, might be an altogether more dangerous place.

# 11

## *Small talk*

As a young missionary, Daniel Everett went to Brazil in 1977
with the aim of converting a remote Brazilian tribe known as
the Pirahã to Christianity. He and his family lived among them
for six years and learned their language. He discovered that
the Pirahã have little sense of time, and no fiction or creation
myths. They live essentially in the present. These aspects of
their lives were an impenetrable barrier to their understanding
of religion, and Everett himself eventually became an atheist.
His interests shifted to linguistics, and he is now dean of arts
and sciences at Bentley University, Massachusetts.

The simplicity of Pirahã life is reflected in their language.
They have no words for colours, and only three words for

counting (roughly translated as 'one', 'two' and 'many').
Their verbs have no tense—no way of signalling the
difference between past and future—other than a simple
distinction between *present* and *not-present*. Their sentences
are simple, without any embedding of clauses. One might be
tempted to think that they suffer from some genetic disorder,
but Everett firmly rejects this.

Everett's account of Pirahã language created a furore among
linguists, because it challenged Noam Chomsky's view that
all humans are born with universal grammar. Chomsky
is well-known for his outspoken criticisms of US foreign
policy, but has also been the dominant figure in theoretical
linguistics for the past fifty years. Universal grammar
implies that languages differ only on the surface, and share
a common, innate underlying structure. Chomsky's main
reasons for this conclusion are that humans are the only
species to have anything resembling language, and that
any child can learn any language. Yet Pirahã language has
very few of the structures one would expect from universal
grammar.

I suspect that the revelations about the Pirahã are part
of a growing realisation that human languages are much
more variable than any notion of universal grammar can
comfortably accommodate. Even the near absence of tense

is not so very unusual. Chinese, for example, has no tenses. Instead, past and future are indicated by the use of adverbs, such as *yesterday* or *tomorrow*, and by what are called aspectual markers, such as the word *before* in a sentence that might be roughly rendered as 'He break his leg before'. In English, too, we can use adverbs, such as *already* and *finally*, and other markers, including calendar and clock times, to specify time more precisely.

Besides insisting that grammar is innate, Chomsky also argues that it did not evolve through natural selection, but was rather the outcome of a fortuitous event, perhaps a mutation, that occurred within the past 100,000 years—and therefore after the emergence of our own species. This view has at least a whiff of creationism, and has been challenged by Steven Pinker in his book *The Language Instinct*. Although he accepts much of Chomsky's linguistics, Pinker argues that a faculty as complex as language must have evolved gradually, shaped by evolution and not by a sudden dramatic mutation. As alluded to earlier, my own view is that language evolved from manual gestures, and can be traced back to grasping behaviour in our primate ancestors.

Much of linguistic theory is based on language as it is written rather than as it is spoken. This may give a false impression of what language is really like, especially in oral cultures. Formal linguistic theory makes much of the way phrases can be embedded within phrases. For example, in the sentence: 'The malt that the rat ate lay in the house that Jack built', the phrase 'that the rat ate' is embedded in 'The malt lay in the house that Jack built'. This is called centre-embedding. Grammar allows further embedding of phrases into embedded phrases, as in the sentence: 'The malt that the rat that the cat killed ate lay in the house that Jack built.' This may take a moment or two to unpack. Add another embedded phrase, and all is lost: 'The malt that the rat that the cat that the dog chased killed ate lay in the house that Jack built.'

There is no such embedding in Pirahã, but even Europeans very rarely use centre-embedding when they speak. Written language is more tolerant of multiple centre-embedding, perhaps because the sentences remain in front of us while we try to process them. The ancient Greeks and Romans may also have been partly to blame. Aristotle laid down the rules for the construction of sentences according to the doctrine of *periods*, where a *periodic* sentence was defined as one with at least one centre-embedding. The Latin scholars Cicero and Livy developed the periodic form and their writing served as stylistic icons for centuries, with a persisting influence in the present-day languages of Europe. But English as she is spoke just ain't like that.

Written language appeared late and is still far from universal. Language evolved to be spoken—or signed, if one happens to be deaf—but not written, and it is shaped much more by culture than by genes. Of course, true language does seem to be restricted to our own species, but I suggest that this is not due to universal grammar, or what Pinker called 'the language instinct' in his book of that title. Rather, it is due to what might be termed a human instinct for inventiveness. We have been extraordinarily successful and creative in moulding our behaviour to suit our environment, and language may be just another case of that.

# 12

## *Music*

Music has been the neglected child of the sciences of the mind. Charles Darwin wrote: 'As neither the enjoyment nor the capacity of producing musical notes are faculties of the least use to man . . . they must be ranked among the most mysterious with which he is endowed.' Steven Pinker declared that musical cognition was not worth studying as it was merely 'auditory cheesecake', a by-product of language.

Certainly, language and music have much in common. Both consist of complex sequences of unlimited variety. Just as there is no limit to the number of different sentences we can produce or understand, so music provides us with a seemingly endless variety of tunes and compositions. Both require precise timing, and both are predominantly based on sound, with a substantial element of bodily movement as well. Language can be accomplished by gesture alone, as in sign languages, and most of us gesture as we speak. Similarly, music gets the body moving through association with dance, or rhythmic tapping.

In many ways, too, language and music are intertwined. The most obvious example is song, where the melodic accompaniment may help to convey emotion. It may also serve a mnemonic function. It is easier to remember words sung to a tune than to remember them as simple sequences, and songs are widely exploited in teaching language to young children. Poetry may also have the rhythmic quality of music, and again may have evolved to help our pre-literate ancestors remember stories and pass them on to the following generations. Ordinary speech itself has a melodic component, known as prosody. The rise and fall of the voice can signal emotion, or differentiate statements from questions or commands.

Many languages, known as tonal languages, use different tones to distinguish between words themselves. In Mandarin Chinese, for example, there are five different tonal contours, so the word approximately rendered as 'ma' can mean five different things. Thus the sentence 'Mama ma ma de ma

ma?', given the right intonation, can be taken to mean
'Is mother scolding the horse's hemp?'—not an everyday
utterance, to be sure, but it makes the point. The language
of the Pirahã, the small Amazon community in Brazil, is also
tonal, and the people there can communicate quite effectively
simply by whistling or humming.

Can you name any note played on the piano without looking
at it? If you can, you have what is known as absolute pitch.
This ability is more common in musicians, especially if
trained from an early age, and is more easily taught to five-
year-olds than to adults. It is fairly rare in our culture, but
common in speakers of tonal languages. This raises the
possibility that speech arose from music, an idea developed
by the archaeologist Steven Mithen in his book *The Singing
Neanderthals*.

The languages with the earliest roots are those of sub-
Saharan Africa, and nearly all of these are tonal. A language
consisting purely of tones would not have the precision of
modern speech, but the addition of supplementary sounds,
including click sounds in Africa and the regular consonants
and differentiated vowels of our own language, would have
ensured greater labelling capacity. In the world's languages,
there are altogether something over 1500 different speech
sounds, but no language uses more than 10 per cent of

them, which could be why tones have dropped out of many languages. Language is over-endowed with phonetic possibility.

Pinker argues that music merely exploits brain circuits that evolved for spoken language, but the reverse may very well be true. Earlier I suggested that language evolved from manual gestures rather than from animal vocalisation, but that need not be incompatible with a musical origin. Both language and music may descend from movement, and rhythm is inherent in most bodily activity, whether in breathing, eating, walking, swimming or, dare I say, copulation. Chimpanzees emit pant hoot calls while drumming their hands or feet against a hard surface, and gorillas beat their chests rhythmically while emitting threatening vocalisations. Therein lies the origin of rock concerts.

If music is auditory cheesecake, then language is muesli, honed from the more holistic processes of dance and melody to provide the communicative precision needed to convey the particularities of our lives, whether social or physical. By the same token, though, it is largely stripped of direct appeal to instinct and emotion. Music still gets in touch with our emotions in a way that speech cannot. As Hans Christian Andersen put it, 'Where words fail, music speaks'.

# 13

## *Remembrance of (some) things past*

Got a bad memory? It's actually much worse than you think, for the simple reason that you don't know how much you have forgotten. I was at a school reunion not so long ago, and was shocked by my failure to remember many of the stories my old classmates told, even though I seemed to feature in many of them. For a while I feared that dementia was setting in, until I discovered that others failed to recognise some of the stories I told. I was consoled by one thing—the memory that some of my old mates were inveterate liars.

Memory, then, is highly selective. Nobody really knows the reasons why some memories stick and others don't. Sigmund

Freud suggested that memories of trauma are repressed, but this theory has not worn especially well. Survivors of the Holocaust, for example, seem to remember even the most horrific experiences. If anything, memories of emotional events are probably remembered better than those of mundane happenings. This is perhaps not surprising from an evolutionary viewpoint, since emotion generally signals happenings that bear, either positively or negatively, on future survival, and remembering emotional events may help us cope differently next time. 'It's a poor sort of memory that only works backwards,' remarks the White Queen in Lewis Carroll's *Through the Looking-Glass*, capturing the point that memory evolved not so much to provide a faithful record of the past as to help one deal with the future.

And even when we do remember past events, they are often remembered wrongly. The unreliability of eyewitness testimony is the bane of courts of law. There may even be survival value in altering our memories, perhaps to boost self-esteem or downplay past embarrassment. Politicians seem especially prone to memory tinkering. In his presidential campaigns, Ronald Reagan often spoke of wartime heroics, but the episodes he described actually came from old movies. Hillary Clinton told of landing in Bosnia in 1996 under sniper fire, but television records show her being greeted in peace by a smiling child. As Mark Twain once said, 'I have been through some terrible things in my life, some of which actually happened.'

American psychologist Elizabeth Loftus, a pioneer of research on false memories, recounts a false memory of her own. She vividly recalls waking up one morning, at the age of fourteen, to find her mother dead in the swimming pool. In her mind she sees and smells cool pine trees, tastes iced tea and sees her mother in her nightgown, floating face down. She cries out in terror, starts screaming, sees the police cars with their lights flashing and the stretcher carrying her mother's body. But she was in fact asleep when her mother's body was discovered, not by her but by her Aunt Pearl. Loftus's memory is a construction, built partly of her knowledge of what happened and partly of extra details supplied by her imagination.

Loftus found that false memories are easily implanted in memory. For example, she asked people to recall events such as being lost in a supermarket, or being taken for a ride in a hot-air balloon, and about a quarter of the people asked gave quite detailed accounts, even though the events had never actually occurred. Memories are also easily altered or implanted by persuasive questioning. In the 1980s some therapists created social mayhem by suggesting to their patients that their symptoms might be due to sexual abuse, often by a parent or close relative who was in fact innocent. This is not to say, of course, that sexual abuse does not occur or that all memories of abuse are false. Nevertheless, the events of that unfortunate decade, along with research by Loftus and others, have led to a better understanding that the probing of memory for past events needs to be undertaken with care, avoiding leading questions and presumptions of guilt. Our brains simply don't work like cameras or tape recorders.

When we speak of memory, we usually mean memory for events or episodes in our personal lives. This is known as episodic memory, and distinguished from semantic memory, which is memory for enduring facts about the world—such as the memory that Wellington is the capital of New Zealand and that sugar tastes sweet. It includes all the words we know, and what they refer to. You probably know something like 50,000 words, along with associated objects, actions, qualities and the like.

In cases of amnesia, episodic memories are typically badly affected, while semantic memories remain more or less intact. One dramatic case is the English musician Clive Wearing, who suffered severe amnesia following a viral infection that destroyed part of his brain known as the hippocampus. Deprived of episodic memory, he lives in the present, constantly under the impression that he has just woken up, or risen from the dead. Yet he can talk, easily recognises his wife and can still play the piano and conduct a choir. His semantic memory appears to be largely intact.

Episodic memories are also profoundly affected in Alzheimer's disease, which may affect up to 50 per cent of people over the age of 85. The disease attacks newer memories rather than older ones, so sufferers often seem to be mentally transported back into earlier periods in their lives. Again, the hippocampus is one of the areas targeted by the disease. But in at least one neurological condition, known as semantic dementia, it is semantic memory that is affected, while episodic memories remain surprisingly unimpaired. If that happens to me, I will have to cease writing essays of this sort, and will probably try to bore you instead with my autobiography. But be reassured, it probably won't work because semantic memory also includes the power of language.

# 14

## *About time*

He said, 'What's time? Leave *Now* for dogs and apes! Man
has Forever.'—*Robert Browning, 'A Grammarian's Funeral'*

Our ability to recall past events is part of what has come to be
called mental time travel. Our memories, however imperfect,
probably evolved not to provide a faithful record of the
past, but rather to supply information for building future
scenarios. This allows us to plan events in detail, or to weigh
up different possibilities before the future is upon us. It also
enables us to create what have been called 'future selves',
models of what we hope to become once we have completed
an education, found a mate, bought a house, chosen a career
or cleaned out the basement.

Like Browning's grammarian, a number of cognitive scientists, myself among them, have controversially claimed that only humans can travel mentally in time. Animal researchers have risen to the challenge, but it has proven peculiarly difficult to show that non-human animals can indeed mentally relive the past or imagine the future. For humans, of course, we can simply ask, but in animals lacking language we need other techniques. The problem is not simple. Consider for example a dog that buries a bone, and later returns to dig it up. This need not mean that it actually remembers burying the bone. It may simply have knowledge of where the bone is buried, without mentally reliving the act itself.

One suggestion is that we could be surer of true remembering if we could show that an animal who buries food can recollect not only *what* is buried and *where* it is buried, but also *when* it was buried. Scrub jays, pesky little birds that cache food for later consumption, may well pass this what-where-when test (also called the www test). If they cache both worms and nuts in different places, they recover the worms when only a short time has elapsed, since they prefer fresh worms to nuts. But if a longer time elapses, they go for the nuts, evidently because they know the worms will no longer be fresh. This suggests that they remember not only what they cached and where they cached it, but also when they cached it.

Moreover, if jays are watched by another while caching food, they later re-cache it, presumably to thwart the watching bird from stealing the food. But they will only re-cache if they themselves have stolen food; even in scrub jays, it takes a thief to know a thief. Re-caching might be taken as evidence that the birds can imagine a future event of theft.

These are among the increasingly clever experiments designed to show that non-human species can travel mentally in time. I'm not yet convinced. My knowledge of such events as my own birth passes the www test—I know where I was born, when I was born and, heaven help me, what was born, but of course I have no actual memory of the event. The critical distinction is between what we know and what we actually remember, and it remains a real challenge to demonstrate that distinction in birds, or in any non-human species. I invite suggestions.

Be that as it may, there can be no doubt that we humans are obsessed with time. Of course, other animals plan for the future by migrating or by hoarding food, but these activities are instinctive, driven by changes of season and have limited provenance. In our own activities we are driven by time itself, with appointments, deadlines, wedding anniversaries and taxes. Although generally adaptive, reliving the past can be an emotional hazard, as when we recall previous embarrassment, and anticipation of the future can cause anxiety, as in contemplating a visit to the dentist or an appointment with the boss—or the inevitability of death, leading perhaps to religions that promise an afterlife. Sometimes, though, the knowledge that time will pass is a comfort, indicated in the proverb 'this too shall pass' or when Viola in Shakespeare's *Twelfth Night*, masquerading as a man, finds herself in an impossible situation and exclaims 'O Time, thou must untangle this, not I! / It is too hard a knot for me t'untie'.

Little is known about how the human brain incorporates a sense of time, but we do know from brain imaging that imagining possible future events activates brain areas that overlap substantially with the areas activated by imagining past episodes. This core network involves the frontal lobes of the brain, as well as the hippocampus, which, as we have seen, is critically involved in memory.

Language itself may have evolved precisely to allow us to share our experiences, and its lack in other species may reflect the absence of mental time travel. Sharing allows us to benefit from the memories and plans of others. And to be adaptive, the tales we tell need not be true. This explains not only our predilection for gossip and shared confidences, but also the human obsession with fiction, whether through stories around the campfire, novels, plays or television soaps.

# 15

## *Coloured days*

The title of Vladimir Nabokov's novel *Ada* carries
a secret code. Nabokov was a synaesthete, who
experienced colours when looking at plain letters of the
alphabet. He saw the letter A as yellow and D as black.
Nabokov was also a lepidopterist who wrote about
butterflies, and the yellow-black-yellow sequence he
saw in the word 'ADA' mimicked the appearance of the
yellow swallowtail.

Synaesthesia means 'joined sensation', in which an item
of one kind elicits a sensation of a different kind. Letters
are normally black printed on white, but the letter-
colour synaesthete nevertheless always attaches the same

colour to a particular letter, and this colour can usually be located precisely on a colour chart.

One way to test for synaesthesia is derived from the Stroop test, a task in which people are shown colour words, such as *red* or *green*, and asked to name the colours of the ink in which they are printed (say blue or yellow). People are slower to respond when the ink and word colours are incongruent (e.g. *blue* printed in red) than when they are congruent (e.g. *green* printed in green). Similarly, if letter-colour synaesthetes are shown letters printed in different colours and asked to name the colours in which they are printed, they are slower to name them when the printed colours don't match the elicited ones. For example, if a synaesthete sees the letter S as green, she is slower to name its colour as red if it's actually printed in red.

By far the most common forms of synaesthesia are colour sensations elicited by letters and digits, or by days of the week. Although colours seem to be the most frequently reported sensations, other elicited sensations include smells, touches, tastes, sounds or temperatures. Similarly, synaesthesia can be elicited by many different kinds of input. For instance, colours have been linked to smells, sounds, tastes—and even orgasm. Curiously, synaesthesia very seldom works both ways. People who associate colours with letters do not see letters when shown colours, and those who have coloured orgasms mercifully do not, as far as I know, have orgasms induced by colours.

In 1880, Victorian scientist Francis Galton estimated that about one in twenty people was a synaesthete. A more recent estimate is one in twenty-three for any type of synaesthesia and one in ninety for those who experience colours induced by letters and digits. One difficulty, however, is to determine exactly who is a synaesthete and who is not. It's partly a matter of degree. For me, Tuesday does have a yellowish-brown tinge, but I don't think I am a true synaesthete. Synaesthesia is also distinguished from metaphor, as in *purple prose* or *hot jazz*. Nor is it a matter of simple association, as in linking yellow with 'banana' or even orange with 'orange'. It is not simply the result of a vivid imagination. Synaesthetic experience is simple, immediate and automatic, with no pictorial elaboration.

Galton observed that synaesthesia tends to run in families, suggesting a genetic influence. This is confirmed in more recent studies, but there are also cases of synaesthesia in just one member of an identical twin pair, implying that a synaesthesia gene, if one exists, is not always expressed. Nabokov's mother Elena was also a synaesthete, as is his son Dmitri. In Elena, though, colours were induced not only by

words but also by musical sounds, suggesting that inherited synaesthesia can take different forms between generations.

In the past, synaesthesia was not taken seriously because of the suspicion that it could be easily faked. However, it gained scientific respectability when imaging showed that areas of the brain corresponding to the synaesthetic sensation were activated. For example, in word-colour synaesthetes the area of the brain responsible for colour processing is activated not only by colours, but also by words alone. The same area is activated by actual colours, but not when people are asked to imagine colours. We therefore now have reason to believe that synaesthesia is indeed real, albeit restricted to a small percentage of the population.

Brain imaging suggests that synaesthesia is due to connections between brain areas, but these connections don't seem to be the result of learning. The colours attached to letters are not those of the letters shown on *Sesame Street* or of those attached to the fridge for young children learning to read. One suggestion is that synaesthesia is a by-product of the way connections are formed in the developing brain.

Newborn babies actually have too many connections in the brain, and these are pruned back in the course of development. Synaesthesia may therefore be the result of incomplete pruning, like a garden whose shoots were not properly pruned back in the spring. It has even been suggested that all newborns are synaesthetes, but most lose it at around three months. This does not explain why synaesthesia so often involves printed words or letters, which are not acquired until after the age of three. Nevertheless, the brain area responsible for processing colour is next to that responsible for identifying words or letters, suggesting the relatively late emergence of reading skills may have invaded an incompletely pruned area already partly dedicated to colour perception.

Synaesthesia, though, is not an infirmity. Besides Nabokov, a number of creative individuals, including Richard Feynman, David Hockney and Jean Sibelius, have been synaesthetes, and one study has shown synaesthetes to have higher average intelligence than non-synaesthetes. In more ways than one, synaesthetes add colour to our lives.

# 16

## *I know what you're thinking*

Much of our social life has to do with figuring out what is going on in the minds of others. Does he love me? Have I offended her? And the lecturer's nightmare: Do they understand what on earth I'm talking about? This awareness of what others might be thinking is known as theory of mind. It underlies our natural tendency to empathise with others, to share their happiness or distress, but it also allows us to interact with others in more complex ways, some of them devious.

Emotion is the easiest state to read, as it is usually written on the face, or in bodily signs. Even mice react more strongly to pain if they perceive pain in others, and monkeys refuse to pull a chain to get food if doing so delivers a shock to a companion. Chimpanzees, but not monkeys, offer consolation to others in distress. A juvenile chimp, for instance, puts a comforting arm around a screaming adult who has been defeated in a fight. Chimpanzees also seem to know when another chimp is looking at them, and steal food when a dominant chimp is not looking.

It is a step up, though, to know what another individual knows or believes, a talent perhaps restricted to humans. One way to assess it in children is the Sally–Anne test. The child is shown a scene involving two dolls, one called Sally and one called Anne. Sally has a basket and Anne has a box. Sally puts a marble in her basket and leaves the scene. While Sally is away, Anne takes the marble out of Sally's basket and puts it in her box. Sally then comes back, and the child is asked where she will look for the marble. Children under the age of four typically say she will look in Anne's box, where the marble actually is. Older children will understand that Sally did not see the marble being shifted, and will correctly say that Sally will look in her basket. They understand that Sally has a false belief.

People with autism seem to lack the ability to read minds. One celebrated case is a woman called Temple Grandin, who has a PhD in animal science and is a professor at Colorado State University. She has written several books, three of which describe her own condition and the manner in which she has dealt with it. In most people, the ability to read minds is instinctive and largely automatic, but those lacking the ability must observe behaviour closely and deliberately to work out what others are thinking or feeling. Grandin has applied this close attention to behaviour to animals as well as to people, which has provided her with information not easily accessible to others. The extra insight gained from this deliberate strategy of observation is reflected in her published research on animals. One of her books (with Catherine Johnson) is aptly titled *Animals in Translation: Using the Mysteries of Autism to Decode Animal Behavior*. A 2006 BBC documentary rather unkindly described her as the woman who thinks like a cow.

Such cases aside, theory of mind can operate to establish intricate networks that guide much of our social activity. In *Twelfth Night*, Maria *foresees* that Sir Toby will eagerly *anticipate* that Olivia will *judge* Malvolio absurdly impertinent to *suppose* that she *wishes* him to *regard* himself as her *preferred* suitor. Each italicised word attributes a state of mind. Theory of mind is recursive—we may fancy we know not only what others are thinking, but also what they think we are thinking. The psychologist David Premack offers the following example: 'Women think that men think that they think that men think that women's orgasm is different.'

A well-known Jewish joke tells of a man who meets a business rival at a train station and asks where he is going. The business rival replies he is going to Minsk. The first man then says, 'You're telling me you're going to Minsk because you want me to think you're going to Pinsk. But I happen to know that you are going to Minsk, so why are you lying to me?'

It has been suggested that theory of mind arose in the Pleistocene, dating from about 2.5 million years ago, when our hunter-gatherer forebears had to bond socially in order to survive on the open savannah, foraging for food and competing with lions, hyenas and other dangerous animals. Nicholas Humphrey has suggested that once they had conquered nature, tribes of hominins began to compete with each other for food, shelter and other necessities of life, leading to the subtle combination of cooperation and deception that drives our social lives today. The dangerous animals we must now deal with are not so much snakes as sellers of snake oil. And they are everywhere—in commerce, politics, religion, and even, dare I say, the university. Take my word for it.

# 17

## *Mirror, mirror on the brain*

Giacomo Rizzolatti is a neuroscientist. With his unruly hair and moustache, he looks like a reincarnation of Albert Einstein. He is excitable, enthusiastic and friendly, and he runs a busy laboratory in Parma, Italy.

Rizzolatti records the activity of single neurons inside the brains of monkeys. In the frontal cortex are neurons which are active whenever the monkey reaches to grasp something, such as a peanut. The recording device can be hooked up to a speaker system, so that whenever a neuron fires you hear a crackling sound. One day, Rizzolatti was surprised to hear this sound, not only when the animal itself reached, but also when a person in front of the animal reached for the nut. Neurons that respond both when the monkey makes a movement and when it observes the same movement made by an individual have come to be known as mirror neurons. They are also described as 'monkey see, monkey do' neurons.

To neuroscientists, this discovery came as a revelation; indeed, one prominent scientist remarked that mirror neurons would do for psychology what deoxyribonucleic acid (DNA) has done for biology. You may well wonder why. The brain has traditionally been viewed as an input-output device, with some neurons responding to particular inputs, such as a face or the call of another member of the species, and some responding when the animal makes some movement, such as grasping something or emitting a cry. Other neurons may lie in between input and output, perhaps representing thought, as when we ponder the meaning of a question before replying. Mirror neurons, though, seem to imply a direct mapping between input to output, as though we have a ready-made system for understanding the bodily actions of others in terms of our own actions. Perhaps you have found yourself squirming on the couch as you watch sport on television, in synchrony with the 'flannelled fools' or 'muddied oafs' on the screen. You can blame your mirror neurons.

Mirror neurons have proved to be quite versatile. In the monkey, they fire not only when the animal sees an action, but also when it hears the sound of a familiar action, such as paper being torn or nuts being cracked open. Brain-imaging experiments show that we humans also have mirror neurons, and they respond to a wide variety of bodily actions. They enable us to resonate with the thoughts and feelings of others, as reflected in their actions, providing a natural explanation for human empathy and the theory of mind discussed in the previous essay. Indeed autism, the condition characterised by an inability to understand the mental states of others, is now widely attributed to a failure of the mirror-neuron system.

Mirror neurons may also help explain how we understand speech, which seems to border on the miraculous. Speech consists of packets of sound delivered at high rates, complicated by the fact that sounds that you hear as the same can actually be very different, depending on the contexts in which they are embedded. For example, *b* sounds in words like *battle, bottle, beer, bug, rabbit, Beelzebub* or *flibbertigibbet* probably sound much the same to you, but the actual acoustic streams created by these *b* sounds varies widely, to the point that they actually have little in common. This surprising fact, discovered only when sophisticated ways of analysing speech sounds were developed, has long been the bane of scientists attempting to programme computers to recognise speech. Computers can now be painstakingly programmed to identify spoken words, but with only moderate success, and nothing like the facility of a normal four-year-old.

One solution to this problem is the so-called motor theory of speech perception, which holds that we perceive speech, not in terms of its acoustic properties, but in terms of how it is produced. The discovery of mirror neurons provided a strong boost for this theory. Thanks to your mirror neurons, you effectively map the speech of others directly onto your own mechanism for producing that speech. Monkey see, monkey do becomes listener hear, listener do.

Mirror neurons almost certainly don't come pre-programmed. They need to be tuned. Much of that tuning probably happens in childhood. My guess is when infants babble, they are in effect tuning themselves to the sounds of speech, and mapping them onto their own production of sounds. That tuning is specific to the sounds of their native language, which is why foreign languages sound not only incomprehensible, but also nearly impossible to parse into their underlying sound patterns. And we are of course tuned into other actions as well, such as sporting skills and playing musical instruments.

It is fitting that neurons were first discovered in the context of arm movements. Rizzolatti gestures enthusiastically as he talks, no doubt activating his own counterparts of the neurons he found in the monkey brain. Mirror neurons could only have been discovered by an Italian.

# 18

## Laughing matters

In 1962, a girls' boarding school in Tanganyika (now Tanzania) had to be closed because of an uncontrollable epidemic of hysterical laughter. Two months later, when the school reopened, the girls started laughing again, forcing a second closure. Scottish poet Carol Ann Duffy's narrative poem 'The Laughter of Stafford Girls' High' describes a similar episode, albeit a fictional one. As these events illustrate, laughter is largely involuntary, and may reduce its victims to helplessness and even, paradoxically, to tears. By the same token, it takes a trained actor to produce convincing laughter on demand, although my father, being Irish, was able to laugh so convincingly at incomprehensible jokes told by a would-be humorist that his mates would later ask what the point of the joke was. He had to confess he had no idea.

Aristotle thought that laughter was uniquely human, but we do find laughter-like behaviour in other animals, especially if they are tickled. Even rats respond to tickling with a sequence of high-frequency sounds that might be taken to be laughter. Laughter also seems to feature naturally in rough-and-tumble play in young animals, involving tickling and non-injurious contact. Chimpanzees at play make a smiley 'play face', with the mouth open, upper teeth covered and lower teeth bared, while they make pant-like laughing noises. You should watch out, though, if the upper teeth are also bared, because this signals aggression rather than playfulness.

Chimpanzee laughter nevertheless differs from human laughter in interesting ways. When we humans laugh we emit sounds in sharp bursts on the outgoing breath, but in chimps the sound is emitted on both the incoming and outgoing breath. In humans, moreover, the sound is rhythmically broken on the same breath, so the sequence of sounds—*ha-ha-ha-ha-ha*—is emitted before you take another breath, whereas chimp laughter remains constant within each breath, and is broken only as the breath switches between incoming and outgoing. Unlike chimps, we humans can alter the shape of laughter, as in *ho-ho-ho*, *ha-ha-ha* or *he-he-he*. In these respects, human laughter seems to reflect the changes in voice control necessary for the production of speech, although laughter itself is very different from articulate speech.

We humans also laugh when tickled. Socially, though, tickling is a ticklish business. You can be tickled by a friend or lover, but tickling by a stranger is generally considered offensive, tantamount to inappropriate touching. Individuals also differ markedly—some enjoy being tickled, some hate it. Tickling can be sexual, and one survey showed that people are seven times more likely to be tickled by someone of the opposite sex than by someone of the same sex.

If you try to tickle yourself it simply fails to register. The reason seems to be that the act of tickling cancels the brain signal that leads to the sensation of tickling. This idea has been tested with a robotic tickle machine, which applies a stroking stimulus to the sole of the foot. This produces a tickling sensation if it operates autonomously, without input from the person being tickled. If the tickled person operates the machine, the tickle sensation is greatly diminished, but increases when a delay is introduced between the hand operation and the movement of the stimulus. Self-tickling with the hand is also more likely to be effective if the hand is on the opposite side of the body from the site that is tickled.

For example, tickling the sole of the foot with the hand on the opposite side is slightly more effective than tickling with the hand on the same side, probably because cross-body tickling introduces a slight delay in the brain between the movement signal and the tickle signal.

Laughter probably originated in play, which is important in the social and physical development of many species. It may have evolved as a signal of non-aggression, allowing play to proceed without threat of injury. Sigmund Freud suggested similarly that laughter is a signal of relief, when an apparent danger proves benign: 'Sometimes,' he said, 'a cigar is just a cigar'—referring, no doubt, to the use of exploding cigars as a practical joke. Slipping on a banana skin is funny only so long as the victim is unhurt.

In humans, then, laughing extends well beyond the slap-and-tickle of play, and we find humour in many situations. But what about other animals? Do turtles have a sense of humour? In the absence of verbal communication it's hard to know, but we can perhaps gain some insight into ape humour from captive chimps or gorillas taught to communicate in signs. These animals do occasionally joke around, as when Koko, a signing gorilla, offers rocks and inedible objects to people as food, or calls her caregiver a 'dirty toilet' when upset with her. More revealing, perhaps, is the male chimp who urinated on his keeper, and then made the sign for 'funny'. Like small children, chimps seem to find hilarity as well as insult in poos and wees.

It's a funny business, laughing, and we still don't fully understand it. But if we did, it might spoil our jokes.

# 19

## *Telling left from right*

After one of his first days at school, our younger son proudly showed us what he had achieved that day. He had printed the entire alphabet, along with the digits from 0 to 9. They were beautifully rendered, but there was just one problem. Each character was written backwards, and the sequence ran from right to left on the page.

Left-right confusion is ubiquitous, especially among young children learning to read and write. The mirror-image letters b and d are often confused, as are p and q, and children may also have difficulty distinguishing words like *was* and *saw* or *dog* and *god*. *Mum* and *dad* remain relatively unambiguous, although *dad* can sometimes be *bad*. Many adults also have difficulty distinguishing their left from their right, as in identifying which hand is which or in giving or receiving directional instructions. As a platoon leader in secondary school cadets, I once gave a sharp 'left turn' command to the platoon, which banged them straight into a cricket sightscreen. I meant to say 'right'. No one was hurt, but I was duly demoted.

In the 1920s and 1930s, the influential American physician Samuel Torrey Orton proposed that left-right confusions were the source of reading disability. As a testimony to this, American author Eileen Simpson wrote *Reversals*, in which she chronicled her lifelong difficulty with left and right, and her struggle with dyslexia. The Orton Society was established internationally to promote Orton's ideas, but it is now recognised that dyslexia can have other causes, and the Orton Society eventually morphed into the International Dyslexia Association.

Left-right confusion is a consequence of a tendency to treat left-right mirror images as the same. This is generally adaptive in the natural world, which is largely indifferent to left and right. Things are just as likely to happen one way round as the other. Predator or prey may lurk on either side, and the same face or body may appear in either left or right profile. Even the letters b and d are really the same, as you can tell by viewing either of them from the other side. It is this indifference to left and right that explains the bilateral symmetry of the body. Our eyes, ears, arms, legs and so on are symmetrically placed so we can be equally aware of what's happening on either side, and respond accordingly. Even the brain is largely symmetrical.

But we need some asymmetry if we are to distinguish left from right. By 'distinguishing left from right' I don't mean simply responding asymmetrically to an asymmetrical input, as when we follow a winding road, or reach for an object on one or other side. I mean the ability to interpret left-right mirror images in symbolically different ways, as in calling the letter b a 'bee' and the letter d a 'dee', or in correctly turning left or right on verbal command. We would be unable to respond in these ways unless we had some built-in asymmetry, which could be as trivial as a mole on one hand, or as complex as an asymmetry in the brain itself.

It is really only in the artificial human world that we must learn to treat left and right as symbolically distinct. In reading most scripts, for example, the orientation of the symbols and the direction in which they run on the page are critical. We must also learn which side of the road to drive on, and some cultures insist that we use the right hand for many social activities, like shaking hands, saluting or eating—

or not to do so, as in a certain toiletry procedure, which is the case in some cultures.

The brain seems to have a built-in tendency to record events both in the orientation in which they actually occurred, and as though mirror-reversed. The reversed version is usually weaker, but may still intrude to create confusion. Patients with brain injury sometimes persistently write backwards, or find it easier to read reversed script than normal script. In one report, a Russian woman with right-sided brain damage drew a map of Russia in reverse.

My fellow researcher Ivan Beale and I suggested that the tendency to treat mirror images, like b and d or p and q, as the same may be due to how the two sides of the brain communicate with each other. Thus if one side of the brain learns that the letter b is a 'bee', it transmits the reverse information to the other side, which then learns that a d is a 'bee'. This reversal has been demonstrated in monkeys and pigeons—for example, if one side of the pigeon brain learns that pecking a right-sloping line (/) brings reward, the other side receives the information that a left-sloping line (\) does the trick. This can further explain why a left-sided brain injury can result in mirror-writing—the right side reverses what the left side has learned. Conversely, in the Russian woman who drew the map in reverse, the left side of the brain may have reversed the map that had been recorded correctly in her now-damaged right brain. In order to avoid left-right confusions in learning to read, then, children may have to establish the dominance of one side of the brain, usually the left, over the other, so that reversed habits don't intrude. Orton recognised that poorly established dominance was one of the factors underlying reading disorders.

Perhaps, then, it is this process of what Beale and I termed interhemispheric reversal that explains our tendency to record events both as they happened and as left-right reversed. This is a nuisance when learning to read, but adaptive in the real world of our forebears. Suppose one was attacked by a dangerous lion from the right, but was lucky enough to escape. If one simultaneously records the event as though the attack came from the left, one would be better off next time regardless of which side the attack came from.

So here's a test for you. You may know James McNeill Whistler's painting of his mother sitting in a chair, but can you be sure whether she is in left or right profile? Or, to take a more familiar example, the Queen's head is shown on all of our coins. Without looking, can you say which profile she is in? Left or right?

# 20

## *The 10 per cent myth*

How often do you hear it said that we use only 10 per cent of
our brains? No one seems to know where this extraordinary
claim came from, although it has certainly been around a
while. A 1929 advertisement for the Pelman Institute, once
famous for its courses on self-improvement, proclaimed that
'Scientists and psychologists tell us we use only about TEN PER
CENT of our brain power', but does not divulge the identity
of the source. Perhaps they had in mind the pioneering
nineteenth-century psychologist William James (brother of
the novelist Henry James), who did once declare that we use
only a small part of our physical and mental resources, but he
seems not to have mentioned a specific percentage.

The suggestion of scientific backing, however spurious, continues to give the 10 per cent myth some leverage in the self-improvement industry. In Steve Biddulph's book *The Secret Life of Men*, the target is men, who are urged to dig further than 10 per cent into their minds and discover the potential for deeper personal relationships. But women are also targeted. Caroline Myss, author of several best-selling books including *Entering the Castle*, appeals to women to go beyond the 10 per cent limitation to find greater powers of intuition and self-fulfilment.

The 10 per cent myth also provides a convenient explanation for extraordinary powers. Albert Einstein is said to have endorsed it, and Uri Geller, famous for his claim to be able to bend spoons through the power of thought, once divulged that he gained his psychic powers by discovering how to break the 10 per cent barrier. Geller's demonstrations are easily duplicated without resort to psychic powers by stage magicians, including James Randi, who wrote a book entitled *The Magic of Uri Geller*—later renamed *The Truth about Uri Geller*. Geller's exploits were also unmasked in New Zealand by two psychologists, David Marks and Richard Kammann, who were able to repeat Geller's demonstrations on television, again without any claim to psychic powers. They too wrote a book exposing the field of psychic phenomena, and Geller in particular, entitled *The Psychology of the Psychic*.

These books are highly recommended, but alas do not have the selling power of books that proclaim the existence of the psychic.

One may even wonder whether we need any brain at all. In 1980, *Science* published an article entitled 'Is your brain really necessary?' This was followed in 1982 by a Yorkshire Television programme with the same title. The provocative question was based on the work of a British paediatrician, John Lorber, who described cases of individuals with hydrocephalus and seemingly massive reductions in brain matter, who nevertheless seemed to function normally. The most striking case was a young man with a measured IQ of 126 and a degree in mathematics, yet brain scans suggested that he hardly possessed any brain at all.

The *Science* article was presented as a news item rather than a scientific report, and the cases described by Lorber were never published as peer-reviewed articles. We don't know whether the young mathematician had normal social skills, or even whether he could tie his shoelaces. In any case, reduced brain volume need not imply a reduction in the actual number of brain cells, which may simply have become packed more tightly. Moreover, the pressure of fluid in hydrocephalus causes brain matter to be pressed against the inside of an enlarged brain case, which can give the illusion of reduced brain matter, just as an inflated balloon may seem to have less physical substance than a deflated one.

The idea that the brain may be unnecessary is especially attractive to those who would like to believe that there is more to the mind than mere matter. Such a belief is comforting, because it suggests that the mind might survive the death of the brain, allowing eternal contemplation of life's mysteries. The vast bulk of evidence, though, speaks to the contrary. Brain damage, no matter how small, nearly always seems to have an effect on mental function, the more so as assessment becomes more sophisticated.

One part of the brain that has posed something of a mystery, though, is the area to the front of the brain known as the prefrontal cortex. One of the most famous cases in neurology is that of Phineas Gage, a railroad construction foreman, who in 1848 was victim of an explosion that saw a metre-long tamping rod blown through the anterior part of his prefrontal cortex, causing major damage. His intellectual function seemed remarkably unimpaired. What did change, though, was his personality. Once responsible and trustworthy, he became irreverent and profane. People with damage to the prefrontal cortex often show very little impairment on routine tests of mental function, but may become disoriented in their everyday lives. The best we can say, perhaps, is that the prefrontal cortex is rather diffuse in its functions. In some ways it may contribute to rather indefinable aspects of our lives, including such elusive concepts as free will.

Modern brain-imaging techniques have allowed us to observe which parts of the brain are active when people are asked to perform mental tasks, from simple word naming to making aesthetic judgements. This work has failed to reveal any part of the brain that is conspicuously silent, as though waiting to be put to use. No doubt we could all have done more with our lives, but we are born to be specialists, with limitations imposed by culture rather than by unused brain space. Each child has the capacity, for instance, to learn any of the world's 6000 or so languages, but no one child could possibly learn more than a tiny fraction of these—even if she uses 100 per cent of her brain. The mental exploits of the human population far exceed those of any individual brain, but most of us probably make a fair stab at using the potential we have. Well, more than 10 per cent, anyway.

# 21

## *Lies and bullshit*

The philosopher Harry Frankfurt once drew attention to a useful distinction between lies and bullshit. When people tell lies, they intentionally state something they know not to be true. One of history's most famous liars was Baron Münchhausen, the eighteenth-century nobleman who told tall tales about his adventures in the German military, including riding on cannon balls and visiting the moon. American presidents seem somewhat prone to telling lies, as when Richard Nixon lied about his involvement in the Watergate scandal, and Bill Clinton denied under oath his relationship with Monica Lewinsky. You might also wonder whether the young George Washington was entirely truthful when he said to his father, 'I cannot tell a lie', after admitting to cutting down a tree. People might even lie about lying.

Bullshit, in contrast, refers to statements made without regard to the facts. Bullshitters don't care whether their utterances are true or false—they just make stuff up. We are continually bombarded with bullshit, from people trying to sell us things, from the media, from everyday gossip. Most advertising and television commercials are bullshit. Many popular ideas, such as homeopathy, Brain Gym, telepathy, much of alternative medicine and perhaps even a fair chunk of orthodox medicine, are largely bullshit.

People are much more censorious when it comes to lying than when it comes to bullshit. The ninth commandment exhorts us not to lie, and children are routinely punished if caught telling lies. George Washington may have calculated that the punishment for cutting down the tree was less than that of having been found to lie. The human species is especially well-equipped to tell lies. We are uniquely blessed with the faculty of language, which provides voluntary control over what we say. In most other species, in contrast, communication is involuntary and fixed. This may have been important for survival, because it means that the signal can be trusted. Crying 'wolf' is effective only if it infallibly signals that a wolf is indeed lurking. There are some exceptions, such as birds that mimic other birds, but human language is more or less unique in that it enables us to weave tangled webs of deceit. This is perhaps why human societies have developed strong sanctions against lying. Without such sanctions, the liar probably always stands to gain, but at the expense of the truthful, and ultimately at the expense of society itself. So society fights back against the liar.

Strangely, though, we seem extraordinarily tolerant of bullshit, to the point that entrepreneurship and promotional exercises are generally admired. Even *Fair Go*, a New Zealand television show normally dedicated to ferreting out liars and cheats, runs an annual competition for the best television commercials, which are judged not for their truth content, but for their entertainment value. It's bullshit that seems to count. Our tolerance for bullshit also means that unscrupulous manipulators can hide their lies under the cover of bullshit. In his poem 'Beware Madam!' Robert Graves warns of the bullshitting lover:

> Beware, madam, of the witty devil,
> The arch intriguer who walks disguised
> In a poet's cloak, his gay tongue oozing evil.

Sometimes, though, we don't really want to know the truth, and prefer the cloak—or perhaps I should say cloaca—of bullshit. We readily accept claims of cures for cancer, potions to prevent ageing, food fads to make us healthier or promises of a paradisiacal afterlife. And of course these things sometimes do have the desired outcome, due to the beneficial effects of belief itself. This is known as the placebo effect, derived from experiments in which neutral substances were prescribed for comparison with an experimental drug. In many cases, saline solution was discovered to be as effective as the drug under scrutiny, or at least more effective than simply doing nothing.

You might think, then, that a placebo shop might be a useful way to make an easy fortune, but it would only work, of course, if the products were not named as placebos. In fact, a great many commercial outlets for placebos, in the guise of alternative medicines or health foods, are already in existence, gumming up the advertising pages and the internet, if not our digestive systems. And they may in fact benefit the gullible, but it is the healthy profit from sales that is the main benefit.

 Unfortunately, bullshit is everywhere. Robert Conquest, in his poem 'A Grouchy Goodnight to the Academic Year', gives this warning:

Then alas for the next generation,
For the pots fairly crackle with thorn.
Where psychology meets education
A terrible bullshit is born.

And remember that the book you have just finished reading was written by a psychologist with a long involvement in education.

# Further reading

Corballis, Michael C. (1993) *The Lopsided Ape: Evolution of the Generative Mind* (Oxford University Press)

———. (2002) *From Hand to Mouth: The Origins of Language* (Princeton University Press)

———. (2011) *The Recursive Mind: The Origins of Human Language, Thought, and Civilization* (Princeton University Press)

Cytowik, Richard E. and Eagleman, David M. (2009) *Wednesday Is Indigo Blue: Discovering the Brain of Synesthesia* (MIT Press)

Della Sala, Sergio (ed.) (2007) *Tall Tales about the Mind and Brain: Separating Fact from Fiction* (Oxford University Press)

Dunbar, Robin (1998) *Grooming, Gossip, and the Evolution of Language* (Harvard University Press)

Everett, Daniel (2008) *Don't Sleep, There are Snakes: Life and Language in the Amazon Jungle* (Pantheon Books)

Frankfurt, Harry G. (2005) *On Bullshit* (Princeton University Press)

Goldacre, Ben (2008) *Bad Science* (Fourth Estate)

Letivin, Daniel J. (2007) *This is Your Brain on Music: The Science of a Human Obsession* (Penguin Books)

Loftus, Elizabeth and Ketcham, Katherine (1994) *The Myth of Repressed Memory: False Memories and Allegations of Sexual Abuse* (St Martin's Press)

Loftus, Elizabeth F. (1996) *Eyewitness Testimony* (Harvard University Press)

McManus, Chris (2002) *Right Hand, Left Hand: The Origins of Asymmetry in Brains, Bodies, Atoms and Cultures* (Weidenfeld & Nicolson)

Pinker, Steven (1995) *The Language Instinct: The New Science of Language and Mind* (Penguin)

——. (2007) *The Stuff of Thought: Language as a Window into Human Nature* (Viking Adult)

Provine, Robert R. (2001) *Laughter: A Scientific Investigation* (Penguin Books)

Rizzolatti, Giacomo and Sinigaglia, Corrado, translated by Frances Anderson (2008) *Mirrors in the Brain: How Our Minds Share Actions, Emotions, and Experience* (Oxford University Press)

Sacks, Oliver (1985) *The Man Who Mistook His Wife for a Hat and Other Clinical Tales* (Touchstone Books)

——. (2010) *The Mind's Eye* (Alfred A. Knopf)

Schacter, Daniel L. (1997) *Searching for Memory: The Brain, the Mind, and the Past* (Basic Books)

——. (2001) *The Seven Sins of Memory: How the Mind Forgets and Remembers* (Houghton Mifflin)

Wearing, Deborah (2005) *Forever Today: A Memoir of Love and Amnesia* (Transworld Publishers)

# Index

Andersen, Hans Christian 59
Aniston, Jennifer 4, 42
apes
    bonobos 7, 11–12, 15, 17
    chimpanzees 7, 9–12, 14–15,
        17, 59, 75, 83, 85–6
    gorillas 9–10, 12, 59, 85
    orangutans 9–10
Aristotle 54, 83
attention 26–30
Auel, Jean 18
autism 75–6, 80

Beale, Ivan 90–1
behaviourism 2
Biddulph, Steve 93
Bilateria 34
bipedalism 11–15, 18
birds 7, 9, 67, 99
    parrots 16
    scrub jays 66–7
    songbirds 16

Bogen, Joseph E. 37–8
brain
    amygdala 3, 47
    asymmetry of 29–30, 33–5,
        38, 89
    colour processing in 72–3
    damage to 28–30, 90, 95
    frontal lobes of 28, 47, 68, 78,
        95–6
    fusiform face area 41
    hippocampus 3, 63–4, 68
    neocortex 9, 48
    size of 7–10, 18, 23
    temporal lobes of 41–2
brain imaging 3, 39, 68, 72, 80, 96
Browning, Robert 65–6
bullshit 97–101

Carroll, Lewis 61
Chamberlain, Lindy 13
Chomsky, Noam 3, 52–3
Cicero 54

Clinton, Bill 97
Clinton, Hillary 61
Conquest, Robert 100
Conradi, Peter 48
Corballis, James Henry 1
Corinth, Lovis 29

Darwin, Charles 56
Denisovans 24–5
deoxyribonucleic acid (DNA) 14,
    23–4, 79
Descartes, René 17, 36
Donne, John 28
Duffy, Carol Ann 82
dyslexia 88

Einstein, Albert 32, 78, 93
Everett, Daniel 51–2

face recognition 40–5
Feynman, Richard 73
Frankfurt, Harry 97
Franklin, Benjamin 32
Freud, Sigmund 60–1, 85
Fry, Stephen 49

Gage, Phineas 95
Gallaudet University 21
Galton, Francis 71
Geller, Uri 93
George VI (king) 48
gestures 17, 19–21, 57
Grandin, Temple 75–6
Graves, Robert 99
Greer, Germaine 50

handedness 31–5
Hebb, Donald O. 3
hemineglect 29–30
Hockney, David 73
hominins 10, 15, 18, 77
    Ardipithecus ramidus 11–12
    Australopithecines 10
    Homo (genus) 10, 12
    Homo sapiens 10, 12, 18–19
    Neanderthals 18–20, 22–5
Humphrey, Nicholas 77
hydrocephalus 94

introspectionism 2

James, William 92
Jerison, Harry 7–8

Kammann, Richard 93

language 9, 14, 16–21, 23, 30–1,
    33–4, 38, 51–9, 64, 66, 68,
    81, 96, 98–9
    gestural origins of 17–21
    sign language 17, 19–21, 57
    speech 16–19, 23, 37, 47–8,
    57–9, 80–1, 84
laughter 82–6
left-right confusion 87–91
Leonardo da Vinci 32
Lewinsky, Monica 97
lies 97–9
Livy 54
Loftus, Elizabeth 62–3
Lorber, John 94

Marks, David 93
McDougall, William 36
memory 60–4
    episodic versus semantic 63–4
    false memories 61–2
mental time travel 65–8
mind reading 74–7
mirror neurons 78–81
Mithen, Steven 58
Münchhausen, Baron 97
music 56–9
Myss, Caroline 93

Nabokov, Vladimir 69, 71, 73
Nixon, Richard 97

Ornstein, Robert E. 38
Orton, Samuel Torrey 88, 91
Orwell, George 12

Picasso, Pablo 32
Pinker, Steven 53, 55, 56, 59
Pirahã, the 51–2, 54, 58
Pitt, Brad 40, 42
Pleistocene, the 77
Premack, David 76

Randi, James 93
Reagan, Ronald 61

Rizzolatti, Giacomo 78, 81

Sacks, Oliver 42
semantic dementia 64
Shakespeare, William 76
Sibelius, Jean 73
Simpson, Eileen 88
Skinner, B. F. 2
speech perception 81
Sperry, Roger W. 37
split brain 36–9
Stroop test 70
swearing 46–50
synaesthesia 69–73

'ten per cent' myth 92–6
Thatcher, Margaret 42–3
throwing 13–14
tickling 83–5
tools 13, 18–19
Twain, Mark 61–2

Vogel, Philip J. 37

Washington, George 97–8
Watson, John B. 2
Wearing, Clive 63–4
Whistler, James McNeill 91